Who Am I?

Australian Animals

Written by Read With You
Center for Language Research and Development

The characters and events portrayed in this book are fictitious. Any similarity to real persons, living or dead, is coincidental and not intended by the author.

Published by Read With You Publishing

Designed by Read With You Center for Language Research and Development

Read With You and associated logos are trademarks and/or registered trademarks of Read With You, LLC.

ISBN-13: 978-1-944710-75-0
ISBN-10: 1-944710-75-2

Third Edition July 2019
Printed in the United States of America.

I only eat eucalyptus leaves.

Who am I?

koala

I can't jump backwards.

Who am I?

kangaroo

I have a backward-opening pouch.

Who am I?

wombat

I store extra fat in my tail.

Who am I?

Tasmanian devil

I have four toes on each foot.

Who am I?

cockatoo

I have two sets of eyelids.

Who am I?

emu

I don't have teeth.

Who am I?

platypus

My loud bird call sounds like human laughter.

Who am I?

kookaburra

Koala
Kangaroo
Wombat
Tasmanian Devil

I am a koala.

I am a kangaroo.

I am a wombat.

I am a Tasmanian devil.

Cockatoo
Emu
Platypus
Kookaburra

I am a cockatoo.

I am an emu.

I am a platypus.

I am a kookaburra.

We are Australian animals.

AMP

Australia has many native animals that cannot be found anywhere else in the world.

Made in United States
Troutdale, OR
05/08/2024